ISRAEL'S ARMOR MIGHT
by SAMUEL M. KATZ

CONCORD PUBLICATIONS COMPANY

Herbert Lugert
Heimbaustr. 7 a, T. 82462
8900 Augsburg 21

Copyright © 1989
by CONCORD PUBLICATIONS CO.
603-609 Castle Peak Road
Kong Nam Industrial Building
10/F, B1, Tsuen Wan
New Territories, Hong Kong

All rights reserved. No part of this publication may be reproduced, stored in a retrieval system or transmitted in any form or by any means, electronic, mechanical, photocopying or otherwise, without the prior written permission of Concord Publications Co.

ISBN 962-361-001-7

Printed in Hong Kong

Front cover

A Syrian tanker's worst nightmare – the business end of a *MERKAVA* Mk. II's 105mm gun. The photograph of the second version of the "Chariot" taken on the Golan Heights perfectly illustrates the *MERKAVA's* unique low slope design; crucial for deflecting anti-tank projectiles on the battlefield.

Back cover

With its turret traversed to re-check already covered ground, a *MERKAVA* crew continues a routine, though never routine patrol along the Golan Heights. Note markings, commander cupola FN MAG equipped with search light, and method for stowing the vehicle's maintenance equipment and crew's personal gear.
(Shmuel Rachmanl /*BAMACHANE*)

4th EDITION

INTRODUCTION

ISRAEL'S ARMOR MIGHT

On 6 June 1982, the Israel Defense Forces launched a massive invasion of Lebanon, and in the process inaugurated a new era in tank warfare. The age of the *MERKAVA* Main Battle Tank was at hand.

The *MERKAVA* or "Chariot" MBT was Israel's answer to years of purchasing its armor fighting vehicles from fickle and sometimes unreliable allies. Conceived in the late 1960s by one of the IDF's master armor strategists, Major General Yisrael Tal, the indigenously produced Israeli MBT was to satisfy the most pressing concern of the IDF Armored Corps: crew survivability. In addition, the tank was to possess considerable firepower, with the least importance placed on speed and mobility which weren't viewed as crucial. The brutal tank battles the IDF's outnumbered armor forces fought atop the Golan Heights against the Syrians, and in the still sands of the Sinai Desert against the Egyptians during the 1973 Yom Kippur War all but demanded fundamental changes in Israel's armor strategy; one a *MERKAVA* MBT could very well satisfy. Although the IDF Armored Corps victoriously fought in some of the largest and closest quarter tank battles since World War Two; heavy, unbearable human losses had been suffered in an army where even a single casualty is one too many. The 1973 fighting also introduced massive and successful deployment of infantry carried anti-tank weapons such as the Soviet RPG-7 and SAGGER; and the deadly toll these inexpensive weapons took on Israeli armor. The obvious need to protect tank crews in battles took on new and urgent importance. In 1979, after years of conceptual design, budgetary concerns and field testing, true production on the "Chariot" MBT began.

The *MERKAVA* was a revolutionary proposal among the world's standard MBT design. A monstrous vehicle weighing over 60,000 kilograms, the *MERKAVA* was the first vehicle of its kind designed with the engine compartment, holding a Teledyne Continental AVDS-1790-6A V-12 diesel, situated up front; protecting the crew from head on fired shells. In addition, the *MERKAVA* was designed with an almost vertical hull front, and well sloped glacis plates to deflect anti-tank rounds. When compared to most MBTs, the *MERKAVA's* crew compartment is spacious, and employs a rear section accessed by a ramp, which can either transport infantrymen or carry additional ammunition. The *MERKAVA's* principle armament was the master of the Middle East battlefield, the proven 105mm gun which had destroyed hundreds of Arab fighting vehicles. In addition, there is a co-axial 7,62mm FN MAG light machine gun, two turret mounted MAGs for anti-aircraft and anti-personnel service, a 50 caliber heavy machine gun mounted at the turret gun base, and a 60mm mortar. The *MERKAVA's* first baptism of fire was the heated battles of the Lebanon madness, where its performance was above all possible IDF expectations, proving to be the finest MBT in the world. Although dozens of *MERKAVAs* were hit and damaged by RPGs, Sagger, and Syrian tank fire; not one *MERKAVA* crewman suffered a fatal injury. The *MERKAVA* even outbattled the most advanced Soviet MBT on the battlefield, the T-72 and its formidable 125mm main gun.

In the years since Lebanon, the Israelis have improved their *MERKAVA* design even further, with a Mk. II variant. According to foreign sources however, the IDF at present possesses fewer than 300 *MERKAVAs*. The Armored Corps as a result has had to rely on a fleet of over 3,000 aging Centurion and Patton MBTs to provide the muscle in its armor punch. In Israeli hands however, these vehicles are far from obsolete. The IDF has always loved to "tinker", and together with its mastery of innovative military thought, and a high tech industry base, it has upgunned and upgraded these "ancient" armor behemoths to serve as front line vehicles well into the next century. Most notable of these inventive modifications are the boxes of "Blazer" or reactive armor fitted to the turrets and chassis of both the Centurion and Patton MBTs. The Blazer armor literally explode on impact, destroying the deadly anti-tank projectile before it can initiate havoc inside the confines of a tank. In Lebanon, Blazer boxes saved countless Centurions and Pattons from destruction. The notion is so successful, that the Soviets have even stolen the reactive armor technology, taken from an M-60 the Syrians captured in Beka'a Valley, and have equipped with their latest armor offering, the T-80.

Supporting the IDF's armor might are thousands of armored fighting vehicles; which ferry troops to the front line, carry field support weapons, and maintain the fighting tanks at the front line. Most prominent among this fleet is the M113 APC, sometimes known as the *ZELDA*. Modified to meet IDF standards with add on armor, and a host of weaponry, the M113 was the workhorse of Israel last major conflict, the 1982 Lebanon War, and promises to be a mainstay of the IDF inventory for years to come. The IDF has also made the maximum use of its "dinosaur" like World War Two vintage M3 halftracks, now outfitted with specially designed rockets to plough a path through enemy minefields. Other armored fighting vehicles, ranging from the tracked monsters deploying self-propelled artillery, to reconnaissance jeeps equipped with machine guns and SAGGERs are considered as invaluable to IDF strategy as the main battle tanks themselves.

In the forty years of Israeli independence, the IDF's Armored Corps *(HEYL SHIRION)* has come a long way since it managed to press into service two stolen Cromwell tanks in 1948. Today, *HEYL SHIRION* is the IDF's strongest combat arm, boasting over 4,000 MBTs, and over 4,000 armored fighting vehicles. Israel's first forty years has resulted in six major wars; with the performance of the Armored Corps a major catalyst for victory in each conflict. Victory however has had a steep price, for the men who wear the black beret. 4,650 tank soldiers have lost their lives in places as diverse and removed as the plush meadows of Latrun, the volcanic Golan plateau, and the urban labyrinth of Beirut. The bitter experiences on the field of battle, and the innovative military thought which is an Israeli exclusivity has bought Israel one dear commodity; the best equipped and most battle-tested tank force in this world.

The next generation of IDF tank solder: Patton crewmen await their graduating inspection at an unidentified base in southern Israel. Their black berets, tank badges and unit insignias are held as dear in their hearts, as the jump wings and red beret are cherished by the paratroops.
(Natan Alpert/*BAMACHANE*)

While its gunner and loader pray on a stop along the road to Beirut, a Patton tank commander examines his vehicle's engine and transmission; June 1982.
(IDF Spokesman)

"Go ahead...make my day"! The business end of the *MERKAVA*'s 105mm gun and low sloped, vertical glacis plates; shown here to advantage.
(IDF Spokesman)

A perfect example of the Israeli tendency to "overload" their APCs with as much as they'll possibly carry! Besides carrying the crew's gear, and additional ammunition, a very convenient method for carrying DRAGON ATGWs has been discovered.
(IDF Spokesman)

With the lead vehicle of a *MERKAVA* Mk. II squad traversing its turret a full 90; an excellent view is afforded of the unique turret design, providing the vital turret area with a maximum of anti-tank protection.
(IDF Spokesman)

The wrong end to be on of a Patton's 105mm gun. Note projection light attached to commander's 7.62mm light machine gun.
(IDF Spokesman)

The IDF's workhorse: the M113 APC (*NAG'MASH*) affectionately known as the *ZELDA*; as seen in its original state, and in its most basic form; transporting reservist paratroopers.
(Yoni Rief/IDF Spokesman)

The true workhorse of the IDF: the M113 *ZELDA* APC; seen here advancing through the streets of Sidon, June 7 1982.
(IDF Spokesman)

Excellent view of the Israeli modified M113's side storage area; holding jerrycans, a rucksack with ammunition, and a spare road wheel. Com-vehicle
(IDF Spokesman)

In operation, and poised against Palestinian targets in the Beka'a valley, a 175mm M107 self-propelled crew pose for the camera. Although the M107 has a maximum range of almost 33 kilometers, the very fact that the gun is located in Lebanon, where friend and foe are one in the same, warrants the flak vests and personal weapons to be close at hand.
(IDF Spokesman)

Pushing his tank's Teledyne Continental ADVS-1790-2A engine to its limit; a Patton commander kicks up the desert sand during mock battles in the Negev Desert.
(IDF Spokesman)

With both the Syrians and Iraqis investing heavily in chemical warfare arsenals as well as in their very public pronounced willingness to use them weapons of mass destruction against Israel; the IDF has been forced to implement immediate counter-measures. During maneuvers in southern Israel, combat engineers prepare to decontaminate a Patton MBT.
(IDF Spokesman)

Beauty and the armor beast! Women armor warfare instructors have been a fundamental element in the success and combat proficiency of the Armored Corps ever since the dark days of the 1973 War. Sitting atop HER vehicle, a *MERKAVA* instructor seems right at home: flak jackets, ballistic helmets, nail polish and all. (IDF Spokesman)

Excellent view of the Centurion APC's concentrated armor configuration, and impressive arsenal of FN MAG 7.62mm light machine guns. Bitter lessons learned at the hands of RPG toting Palestinian and Shi'ite guerrillas have taught the IDF the necessity for a "merkava type" crew survivable personnel carrier. (Herzl Kunesari/IDF SPOKESMAN)

Excellent view of the nose attachment mechanism which can accommodate a wide array of equipment; from obstacle removing ploughs to bulldozer blades. (IDF Spokesman)

Close-up of the M113 communications variant; with its stowage area seen to advantage. (IDF Spokesman)

A modified Patton on the "prowl"; hunting enemy *MERKAVAs* during battle training in the Negev Desert. (IDF Spokesman)

A side view of an upgrade Patton; with its rear turret stowage area, and turret gun base .50 caliber machine gun mounting seen to advantage.
(IDF Spokesman)

Advancing towards Beirut in haste, a Patton fitted with dozens of Blazer reactive armor boxes shows to advantage the differing shapes, sizes, and conformoties of this ingenious development in crew and vehicle protection.
(*BAMACHANE*)

Close-up rear view of the *MERKAVA*. Note interior construction of rear turret storage rack.
(Author's Collection)

Excellent side view of the tracks, road wheels, and suspension apparatus of a *MERKAVA* Mk. II, covered with a few layers of dust, sand, and "experience" after a successful winter's maneuvers in the Negev Desert.
(Herzl Kunesari/IDF Spokesman)

An M109A1 gun in operation against Syrian forces in the Beka'a valley, during the initial Syrian-Israeli battles of the 1982 War. Note ready rounds near vehicle rear.
(IDF Spokesman)

At a paratroop and infantry exhibition in the red beret's adopted city of Ramat Gan, a field ambulance stands on guard for any contingency. Note large *MAGEN DAVID* (Star of David) indicating a medical vehicle.
(Samuel M. Katz)

The cramped, sometimes claustrophobic interior of the M113 *ZELDA*. The exterior fuel tanks are seen here to advantage.
(Sigalit Katz)

After receiving a "gift from the sky"; paratroopers quickly unload their M113 from its airborne carrier during a training maneuver in central Israel.
(IDF Spokesman)

An impressive view of an impressive fighting vehicle; as seen on the flat volcanic plateau of the Golan Heights. Note protective manner in which the front portion of the side skirt armor protects the road wheels.
(IDF Spokesman)

A modified Patton armed with the KMT-4 I.M.I. anti-mine clearing device. Note rear ammunition carriage towed in line.
(IDF Spokesman)

This close-up view of the *MERKAVA* Mk. II offers an excellent view of the additional armor plating welded to the turret. Note Type 602 ballistic helmet and its FN MAG stock resting place.
(IDF Spokesman)

Excellent rear view of two upgraded Pattons racing in for the kill! Note manner in which the last vehicle; fitted the mine clearing roller, succeeds in negotiating the most difficult of terrain with relative ease.
(IDF Spokesman)

Originally the M163 VADS 20mm Vulcan Gatling autocannon with ranging radar was obtained by the Israel Air Force (IAF) as a low level, mobile air defense weapon. Its performance in Lebanon however, where its rapid firing gun proved devastating against enemy fortifications, earned it the distinction of an infantry support role. Like all other "imported" vehicles in the IDF inventory, the M163 has been subject to a fair share of modifications, as seen in these two photographs such as the side stowage racks.
(IDF Spokesman)

The importance given to individual tank anti-aircraft and anti-personnel capabilities are perfectly illustrated here; by the two turret mounted FN MAG 7.62mm light machine guns, and the .50 caliber gun mounted at the main armament, M68 series 105mm gun turret base.
(IDF Spokesman)

Not a good place to have a tread breakdown; and even a worse one to take a breather: a Patton crew awaits the answer of its maintenance S.O.S. in southern Lebanon, August 1984.
(IDF Spokesman)

On patrol against Palestinian and Shi'ite terrorists, the crew of a *MERKAVA* MBT nicknamed *MAHATZ* (Hebrew lettering combined with unit designation on side skirt) or "Strike" carefully scours the southern Lebanese countryside for "bandits" in 1983. The crewmen wear the Orlite Type 602 Kevlar ballistic helmets.
(IDF Spokesman)

Although this side view does not do justice to the *MERKAVA*s well sloped glacis plate, its unique design, meant to offer AT rounds the minimum angle of impact is clearly evident.
(IDF Spokesman)

Excellent view of the *MERKAVA* Mk. II's road wheels, and the M68 105mm gun's thermal sleeve for 105mm gun.
(Herzl Kunesari/IDF Spokesman)

What the M113 is to IDF mechanized strategy; the M109A1 155mm self-propelled artillery piece is to the Artillery Corps. The corps' workhorse, Lebanon was its first true test as a long range support gun; as well as an effective close-range support weapon. This M109A1, seen here along the road to Beirut in June 1982 sports its typical IDF trademarks, especially the crew's gear stored on the modified stowage bins. Note .50 caliber M2 machine gun fitted above the main gun; a very useful addition when spotting at close ranges; especially in bandit country!
(IDF Spokesman)

A far cry from what the original British designers intended it to look like: a HEAVILY upgraded Centurion MBT produces a cloud of sand during desert exercises. Note boxes of Blazer armor, improved commander and loader cupolas, thermal sleeve for 105mm gun, and smoke grenade discharger. Note camouflage netting "neatly" folded by turret rear.
(IDF Spokesman)

The dried desert flowers offer a magnificent perspective of the *MERKAVA*'s unique low slope design; seen here in the Beka'a Valley during a battle with Syrian forces, June 1982.
(IDF Spokesman)

At a unit rest stop along the road to Beirut; a *MERKAVA* Mk. I MBT and its battle-exhausted crew take a breather. Note orange aerial identification sheet crudely folded stop the turret.
(IDF Spokesman)

The elevated 105mm gun of this *MERKAVA* on the Golan Heights allows the turret gun base attachment for the .50 caliber machine gun to be seen to advantage.
(Yoni Reif/IDF Spokesman)

Immersed in enough mud to become a crew's nightmare; a modified Patton M60A1 stands guard at a base in northern Israel. Note rear turret storage attachments; and KMT-4 anti-mine roller device.
(*BAMACHANE*)

Amid the tall burnt grass of the Golan Heights; three M113s bristling with antennas and weaponry conduct a routine patrol.
(IDF Spokesman)

An excellent view of a *MERKAVA* force deploying for battle, supported by an M113 ferrying *GOLANI* infantrymen; and a command jeep.
(IDF Spokesman)

An M113 crewman, wearing a snow suit, flak jacket and Type 601 ballistic helmet observes the "seemingly" peaceful terrain of southern Lebanon.
(IDF Spokesman)

To counter the enormous volume of Soviet anti-tank mines in the Arab arsenal, Israel Military Industries (I.M.I.) has developed several indigenous responses. Leading a company charge during large-scale maneuvers, a *MERKAVA* Mk. II fitted with the KMT-4 mine detonation rolling device ploughs through a "minefield".
(Herzl Kunesari/IDF Spokesman)

A TOW fitted M113 during night time firing exercises. Of interesting note is the *MAPATZ* tri-pod for infantrymen use, attached to vehicle front.
(Jonathan Torgovnik/IDF Spokesman)

A heavily modified Patton in the mud of northern Israel. Note thermal sleeves attached to 105mm gun, and small boxes of Blazer armor attached to vehicle nose.
(*BAMACHANE*)

An upgrade M-60 Patton MBT equipped with boxes of Blazer reactive armor prepares to cross a water-obstacle during company exercises.
(IDF Spokesman)

The weapon to counter the vast amount of SCUDs and FROGs in the Arab arsenal; the U.S. made LANCE surface-surface missile.
(Sigalit Katz)

In the barren emptiness of the Negev Desert; a *MERKAVA* Mk. II prepares to set itself down from its tank transport and head out for the next destination on the exercise itinerary. Photograph offers interesting view of rear turret "ball and chain" AT round deflection mechanism, and commander copular mounted FN MAG 7.62mm light machine gun.
(Hanoch Gutman/*BAMACHANE*)

MECHANIZED WARFARE ISRAELI STYLE! A column of M113s, equipped with a wide assortment of weaponry including mortars and machine guns; races through the desert sands.
(No'am Armon/*BAMACHANE*)

Following in close pursuit to their lead Patton tank, paratroopers give all they got during large-scale maneuvers in the Negev Desert.
(IDF Spokesman)

Old reliable! One of the IDF's most enduring artillery pieces of both the 1973 and Lebanon Wars, the M109A2 155mm self-propelled artillery piece. Note black boards used to indicate vehicle markings; and gun's nickname "Granite" stenciled on the gun.
(Michael Giladi/IDF Spokesman)

All set for some 50 caliber target practice, a ZELDA crew prepares for the firing order.
(Jonathan Torgovnik/IDF Spokesman)

Kicking up a cloud of dust, a MERKAVA MBT races into firing position in southern Lebanon.
(IDF Spokesman)

At the rehearsals for the IDF's 40th anniversary parade, the two other facets of the IDF's mechanized anti-aircraft effort sit in patience: the mobile HAWK SAM, and the M48A1 Chaparral.
(Sigalit Katz)

Getting his *MERKAVA* Mk. II into position for a night patrol in "bandit country", a flak-jacked tank commander negotiates a minor obstacles and displays his vehicle's impressive terrain handling qualities. Note 50 caliber machine mounted on the turret's 105mm gun base.
(Yasi Tzevker/BAMACHANE)

A *MERKAVA* Mk. II MBT named "Golan" offers an interesting side view of the famed "chariot tank", including heavily armored side skirts, seen here with reinforced welded plates.
(IDF Spokesman)

Although not designed for its speed and mobility, a *MERKAVA* Mk. II kicks up some desert sand as it races into firing position. Note additional armor plating on turret, and orange aerial identification sheet, a relic of the Lebanon period.
(Courtesy Herzl Kunesarl/IDF Spokesman)

Enveloped in a cloud of smoke and dust, a Centurion continues its advance during live firing training exercises. Note add-on plate of armor attached to glacis plate.
(Ofer Karni/IDF Spokeman)

Amid the beauty and alleged serenity of a Lebanese forest, a Centurion *(SHO'T)* takes aim and fires at retreating Syrian commandos in the Shouf Mountains; July 1982.
(BAMACHANE)

A *MERKAVA* MK. 1 MBT crew enjoys a much welcomed departure from combat duty in southern Lebanon, Spring 1985. The large Star of David flag displayed on the vehicle's aerial was seen much more in friendly territory; than in areas known to be inhabited by Palestinian and Shi'ite guerrillas.
(Michael Zarfati/BAMACHANE)

The IDF's answer to enemy obstacles: the M48/M60 AVLB (Armored Vehicle Launched Bridge).
(Michael Giladi/IDF Spokesman)

Equipped with a bulldozing blade for some "obstacle removal", a cautious tank commander leads his upgraded Patton across an M48/M60 AVLB.
(Shlomoh Arad/BAMACHANE)

Few armies in this world push their men and equipment into such diverse terrain conditions as the IDF: having the same vehicle perform mobility exercises in the desert, and then having it patrol a vital mountain border post in Alp like snow near Mt. Hermon. (*BAMACHANE*)

Rear view of a modified M113, with its power operated rear ramp door opened. Note the unique construction of the storage bins; which allow easy storage access, yet keep the load secure during movement. (Sigalit Katz)

The success of the Artillery Corps in recent combat has been as a result of the determination, and dedication of the battery crews; typified here by the grim zeal this gunner puts in his effort.
(BAMACHANE)

Kicking up a cloud of smoke and a host of salutes, a column of Pattons parade for inspection before the IDF *KA'SHNA'R*; Chief Armor Officer during ceremonies in sourthern Israel.
(Mike Mirhaim/IDF Spokesman)

The IAF's stop gap low level AA weapon: the M48A1 Chaparral. Its effective altitude range of 50 to 3,000 meters has made it ideal to counter low flying bombers, as well as the new nemesis to armored vehicles proved so deadly in Lebanon: the helicopter gunship.
(IDF Spokesman)

On the war scorched earth of a Lebanese field, two *MERKAVA*s await their order to fire against Syrian positions in eastern Lebanon; June 1982.
(IDF Spokesman)

A *MERKAVA* sitting in defensive position on the Golan Heights with its turret shown to advantage. Note fume extractor located near commander's cupola mounted FN MAG 7.62mm machine gun.
(IDF Spokesman)

A derivative of the IDF's love for tinkering, and its need for heavily armored personnel carriers: the Centurion APC. Upgraded with additional armor plating; including boxes of Blazer reactive armor, the Centurion APC is heavily armed with 50 caliber and 7.62mm machine guns. This vehicle has been modified to the specifications of the Combat Engineers (*HEYL HANDASAH*); as a sapper "shock troop" vehicle.
(Michael Giladi/IDF Spokesman)

Beautiful photograph of a jeep mounting a TOW ATGW; racing past rows of M113s just after the capture of Tyre, 7 June 1982. The TOW jeeps, usually deployed by reconnaissance paratroop or infantry units has proved to be a most lethal anti-tank combination: combining speed and accurate firepower. Note missiles crudely fastened near driver.
(IDF Spokesman)

Another relic of days past: the ubiquitous M3 halftrack; fitted here with dual FN MAGs and an improvised canvas roof for security duty along Israel's frontiers.
(IDF Spokesman)

The *MERKAVA* was designed for crew survivability and comfort; a concept expressed in the revolutionary web gear issued to IDF tank crews consisting of modified chest pouches.
(Herzl Kunesari/IDF Spokesman)

Lined up in defensive position in the coveted "high ground", a squad of upgraded Pattons equipped with obstacle clearing devices attached to the vehicle nose, await the anticipated order of *ESH*: FIRE!
(Jonathan Rief/IDF Spokesman)

Brigadier General Yossi Ben-Hanan, the OC IDF Armored Corps, salutes the tank commander of a *MERKAVA* Mk. II during an inter-service ceremony. Note the *MERKAVA*'s distinctive pointed turret front.
(Natan Alpert/*BAMACHANE*)

With the invaluable sleeping bags securely fastened, a Patton races towards its objective in haste.
(IDF Spokesman)

Deployment in the desert! An upgraded M113 armed with FN MAGs and an "out of place" 30 caliber machine gun get into attack position during combined arms exercises. Note unit insignia stenciled in white on vehicle front.
(*BAMACHANE*)

On patrol in southern Lebanon, an M113 communications vehicle passes through the rolling hills; which although seem peaceful at the moment; can host a wide assortment of adversaries. Note how glacis plate area of vehicle has been turned into a virtual storage compartment, holding crates of ammunition and the crew's personal gear.
(IDF Spokesman)

An M579; the repair version of the M113A1 in a Lebanese citrus grove; near the bitter close quarter fighting at the Ein el-Hilweh refugee camp. The M579 is fitted with a hydraulic crane on the roof for repair of light armored vehicles.
(IDF Spokesman)

As realistic a battlefield simulation as one can get, an upgraded Patton races through the desert terrain, and the spectacular effects of an exploding phosphorus shell.
(*BAMACHANE*)

A heavily armed M113 conducts a routine patrol of an unidentified village in southern Lebanon. Note two FN MAGs, .50 caliber machine gun; and 60mm mortar.
(Gilad Shekma/*BAMACHANE*)

The occasion of former Chief of Staff Lt. General Moshe Levy (left) congratulation an exemplary *MERKAVA* crew poised opposite Syrian forces in Lebanon 1983 offers a good view of the vehicle's side skirt armor construction. Note 105mm gun's -8.5 depression; seen here to advantage. The tank commander's (center with field glasses) flak jacket was a Lebanon necessity.
(IDF Spokesman)

The IDF Armored Corps learned the invaluable defensive advantage of "hull down firing positions" during the Syrian blitzkrieg across the Golan Heights on Yom Kippur Day, 1973. That lesson; learned and developed to its maximum potential is seen here in practice on the Golan Heights.
(IDF Spokesman)

An ingenious response to a turret's anti-aircraft and anti-personnel capabilities. Twin mounted 50 caliber machine guns, and their improvised jerry can ammunition storage bins.
(Mike Mayrhaim/IDF Spokesman)

A break during an exercise offers the crew a valued few moments of relaxation. Note load Patton's commander cupola mounted FN MAG with attached ORTEK ORT-T2 3.5x passive night sighting system.
(IDF Spokesman)

Centurions in battle! Assisting paratroop elements during the 4 May retaliatory raid against Shiíte terrorists in the south Lebanon village of Maidun; a column of upgraded Centurions await their call to action. An excellent view is afforded of the application of the Blazer reactive armor applied to the Centurion MBT. (Natan Alpert/*BAMACHANE*)

Excellent view of the rear turret area, with its ball and chain AT round deflection system and very large storage rack shown to advantage. Note folded stretcher (right) carried near rear crew hatch. (IDF Spokesman)

Close-up view of the modified stowage racks the IDF Ordinance Corps has experimented with; providing crews not only with additional cargo space; but added protection as well.
(IDF Spokesman)

At a company staging point, a modified Patton with a roller type mine clearing device prepares to clear an attack path for an M113 ferrying paratroopers into mock battle.
(IDF Spokesman)

Although it is a hulking mass of steel; weighing over 60 tons when equipped for battle; the *MERKAVA* is a very agile MBT. Note design of wavy bottom side skirt armor.
(IDF Spokesman)

The 20th century's version of the cavalry charge of yesteryear: *MERKAVA* Mk. IIs practice an attack formation, as they kick up the parched earth of a northern Israel field.
(IDF Spokesman)

A "lucky *MERKAVA* crew" not only gets the opportunity to have former Chief of Staff Lt. General Moshe Levy ride with it; but is fortunate enough to have their division commander personally directing their every move. Note exterior fire extinguisher attached near unit markings.
(IDF Spokesman)

Truly the master of desert warfare. Prepared for a truly long stay in the barren Negev, an upgraded Patton conducts a routine patrol. Note sleeping bags attached to rear turret area.
(IDF Spokesman)

Performing its sacred task of punching a hole through enemy defenses for the infantry to flow through, a *MERKAVA* leads two M113s filled with paratroopers during desert exercises.
(IDF Spokesman)

Intimidating just by their monstrous appearance, two *MERKAVA*s practice combat patrols in the wide open spaces of the Judean Desert.
(Herzl kunesari/IDF Spokesman)

Night patrol preparations for a paratroop unit in southern Lebanon. Rear angle offers an excellent view of add-on armor plates; and the improvised method of stowing gear inside them.
(Gilad Shekma/*BAMACHANE*)

Seeking cover by a lone desert tree, a modified Patton attempts to evade "enemy" Centurion tanks and Cobra helicopter gunships. Note orange aerial identification sheet draped over rear turret storage areas.
(IDF Spokesman)

"The most effective anti-sniper weapon ever invented": the M163 20mm VUI-CAN SPAAG (Self-Propelled Anti-Aircraft Gun). Originally obtained by the Israel Air Force (IAF) as a field unit answer to low flying jet and helicopter threats, the "Vulcans" proved their multi-faced value during the 1982 invasion of Lebanon when their rapid firing 20mm gatling guns proved devastating effectiveness against Palestinian fortifications, and especially in rooting out snipers! (Michael Zarfari/*MABMACHANE*)

Side view of an upgraded Patton, showing the unique shapes and sizes of its customized layers of Blazer reactive armor; designed specifically to surround the entire turret.
(Herzl Kunesari/IDF Spokesman)

Fitted with an anti-obstacle device, a Patton carefully glides through the snow of south-eastern Lebanon. Note elevating size of Blazer armor which surrounds turret area.
(Mike Mayrhaim/IDF Spokesman)

At a maintenance lot at a base in northern Israel, crews check the interior and exterior of their *MERKAVA* MBTs.
(IDF Spokesman)

Like any weapons system the IDF imported from abroad; the M113 was not spared Ordinance Corps modifications. The most notable and perhaps most typical are the add-on armor sections which surround the vehicle's main frame. As perfectly illustrated in the photograph, the sections are functional, and provide easy access to the original vehicle. The added protection, as perfectly expressed by the smiling faces of this *ZELDA*'s crew, is a much welcomed commodity.
(IDF Spokesman)

Rolling through the desert hills in haste, a Patton named *SE'ARAH* ("Storm") makes its way towards its parent battalion during brigade size maneuvers. Note manner in which Blazer armor completely covers the glacis plate, and obstacle clearing device attachment on the vehicle's nose.
(IDF Spokesman)

An ancient battlefield and a rookie MBT which has already seen its share of combat: *MERKAVA*s patrol the very volatile Golan Heights.
(IDF Spokesman)

A truly mechanized way to go war! Two GOLANI Brigade M113s (including an ambulance identified by its white circle and red Star of David) lead their armor support into the fray.
(IDF Spokesman)

On a firing field in northern Israel, two *MERKAVA*'s patiently await their turn to destroy a target or two during battalion exercises. Note ball and chain AT round deflection devices attached to rear turret area.
(IDF Spokesman)

At a defensive position "somewhere" in northern Israel, an upgraded Centurion mans the vigil. Note absence of traditional Centurion side skirt armor, Blazer armor applied to vehicle glacis plate, and turret; and smoke grenade dischargers alongside the main armament 105mm gun complete with thermal sleeve. To enhance its frontal armor protection; an obstacle removal device complete with spare road wheel has been attached to the tank's nose.
(IDF Spokesman)

Note add-on skirts of armor complimenting the IDF's modified M113 APCs. By transforming a 30 year old armored vehicle into one able to withstand today's saturated battlefield environment, the casualty and budget conscience IDF has been able to extend the service life of these APCs into the next century. (Michael Giladi/IDF Spokesman)

A frontal view of the IDF M163. The accurate, fire controlled multi-barrel 20mm cannon has enabled a single vehicle attached to a small unit to saturate an area with heavy fire power.
(IDF Spokesman)

This convoy of M113s along a "secure" road in southern Labanon offers a unique view as to the extensive modifications the IDF has applied to its beloved *ZELDA*. The first two APCs have been fitted with additional crew storage space to hold sleeping bags, stretchers and additional ammunition; while the vehicles bringing up the rear have been upgraded with additional sheets of protective armor.
(IDF Spokesman)

Although the add on skirts of armor do not "soften" an AT round's angle of impact on a vehicle; it does diminish the penetration ability of such weapons as an RPG and SAGGER, and most importantly, provides additional protection to crews.
(IDF Spokesman)

A close-up view of the Blazer reactive armor applied to the Patton turret, as well as the unique swing around gun base for the loader's FN MAG 7.62mm light machine gun; offering the crewman a virtual 180 firing capabilities.
(IDF Spokesman)

Enjoying a quick bite; a Patton crew take a breather atop their vehicle's ABK-3 add-on bulldozer blade attachment produced by Israel Military Industries. Note modified rear turret storage bin area, and .30 caliber machine guns replacing the normal turret mounted FN MAGs.
(IDF Spokesman)

Excellent view of the extra skirts of armor which now surround most IDF M113s. Note crew in full NBC warfare protection gear.
(IDF Spokesman)

View of an upgraded Centurion (know in IDF slang as a *SHÓT*); once the mainstay of the IDF Armored Corps, on tank transporters heading into the Lebanon fray: June 1982.
(IDF Spokesman)